# Non-Propellant Propulsion for Space Flight

# Non-Propellant Propulsion for Space Flight

Yung-Kang (Derby) Sun, Ph.D.
Engineering Design and Development Group

ISBN 978-1-300-88534-4

# Contents

# Acknowledgement

First, I would like to thank Mr. Sean McBride for his generous assistance in my research during the discussions. Next, I would like to thank Dr. Ray Wimberly, senior lecturer in the Department of Mechanical and Aerospace Engineering at the University of Texas at Arlington, for his variable technological advises in this unique research. Also, I would like to appreciate Mr. Rolando Castilleja and Mr. Kurata Hironari for their technical supports to transfer the concept into practical applications. In addition, I would like to thank the following generous supporters from Kickstarter Incorporated to help me publish this book:

S Lyall Hunt
Theodore Martin
David Ells
Eric Moeller
Stephen Rider
Andreas Flato
Don Robertson
Stephen Rubin
David Churn
Damian Gordon
Matt Selter
Kevin Sue
Jochen Schmiedbauer
Nicholas Bevan

At this end of this book, I would dedicate to my mother, Li-Yun Hung, for her unconditional support from my beginning of this research and academic studies.

# Preface

Our ancestors are curious other unknown worlds when staring to the sky at dark nights. Like Captain James T. Kirk, the fictitious character in the Star Trek created by Gene Roddenberry said, "Space, the final frontier. These are the voyages of the starship Enterprise. Its five-year: to explore strange new worlds, to seek out new life and new civilizations, to boldly go where no man has gone before", they always dreamed to travel to unknown environments and meet domestic entities other than Earth someday. Since the beginning of the Space Age, human beings have reached marvelous achievements to send astronauts to the Moon, and probes to other planets far away from out home planet. When George W. Bush, Jr., the 43$^{rd}$ President of the United States, unveiled the new vision to put human beings back to the Moon, and explore to the Mars and beyond in 2004, scientists and engineering researchers from academic and industrial institutions, and governmental agencies proposed various creative conceptual theories and designs to respond this challenge from the national leader. During the Apollo mission in the 1960s, the Service Propulsion System in the Service Module played the key role to propel the Lunar Module with three astronauts, limited scientific equipment and life supplies from the Earth's orbit into the Lunar orbit for three days. According to the design revealed from NASA, the Service Module carried 40,590-pound propellant. The propellants are highly explosive chemical components in pressure tanks to operate under the harsh and uncertain environment in space. The total percentage of the propellant weight in the Service Module is around 75.08% compared to the total weight of the whole Service Module that is 54,060 pounds! In the Apollo 13 mission, the dangerous propellant exploded on the way to the Moon without any explanation. Fortunately, all three crewmembers flied back home safely from the great assistance in the Mission Control Center at NASA Johnson Space Center. Since national leaders around the world expected to corporate space missions to other planets, the safety for astronauts and reliability of the space propulsive system are main issues for planetary travels under dangerous conditions, such as radiation and plasma in space. In order to provide economical and reliable propulsive technologies for interplanetary transportation under the cruel world economy, scientists and engineers must develop unique propulsive systems required less or no propellant other than conventional propulsive devices to transfer mission crews from Earth to target planets for scientific explorations. Because no advanced propulsive technology is under the test in real space missions at this moment, this does not mean human races could never have any chance to visit other territories and discover other life forms forever. As a member in the space engineering community, I would be glad to share my previous and current work regarding to unique space propulsion systems published in different conferences held by the American Institute of Aeronautics and Astronautics (AIAA) from the year of 2005 until now in this book. Like Dr. Robert H. Goddard, the pioneer of the rocketry once said, "this is difficult to say what is impossible; for the dream of yesterday is the hope of today and the reality of tomorrow". Because of dreams for space explorations, we always put the great hope to develop new propulsion technologies to propel space carriers to other planets that make space explorations become the reality.

In conclusion, I would like to express my sincere appreciations to scientists and engineers whose excellent comments and suggestions were both helpful and welcome from academic institutions and research agencies.

Yung-Kang (Derby) Sun, Ph.D.

# Chapter 1

## Introduction

Non-propellant propulsion is a unique conceptual technology in space propulsion for interplanetary transportation. From 1996 to 2002, Marc G. Mills, a leading expert on Breakthrough Propulsion Physics at NASA John H. Glenn Research Center set the new requirements in search of creative innovations in space transportation. The requirements of innovative propulsion are:

1. The breakthrough propulsion system does not carry any propellant mass.

2. The breakthrough propulsion system reaches the maximum possible transit needs during interplanetary flight.

3. The breakthrough propulsion uses the power provided from the spacecraft during interplanetary flight.

Based on the requirements, topics of interest include experiments and theories regarding the coupling of gravity and electromagnetism, the quantum vacuum, hyper-power fast travel, and super luminal effect. Because the goals of propulsion are presumably far from fruition, a special emphasis is to identify affordable, near-term, and credible research that could make measurable progress toward the purposes of practical applications in advanced propulsion.

In the standard procedure for planetary exploration at the Apollo Program, the Saturn V rocket, the giant launch vehicle in the history, must deliver the Apollo Spacecraft from the ground to the appropriate ready parking orbit. When approaching the injection position, the Service Module fired the engine to propel the Lunar Module and Command Module with three astronauts and restricted weight of life supplies and scientific equipment that are 66871 pounds (30,324 kilogram) to the lunar orbit for three days. Figure 1 shows the process of the lunar flight for the Apollo mission in the 1960s. Although NASA and other space agencies around the world prefer to transfer mission crews from the Space Station orbiting on Earth in order to reduce complicated procedures similar to the Apollo era, propulsion is the major challenge to send space explorers to other planets in the solar system. According to the NASA's evaluation, the new space vehicle still needs three days to carry astronauts from the Earth's orbit to the Moon, and three months to Mars with advanced chemical and nuclear propulsions under appropriate orbit transfer methods. Mission crews would face heath issues under highly radioactive environment in space and limited room in the spacecraft during the interplanetary flight for months. In order to increase propulsion capacity for space vehicles in the short-time span, engineers and scientists must develop new propulsive theories and technologies with creativity and imagination based on firm foundations under the classical physics.

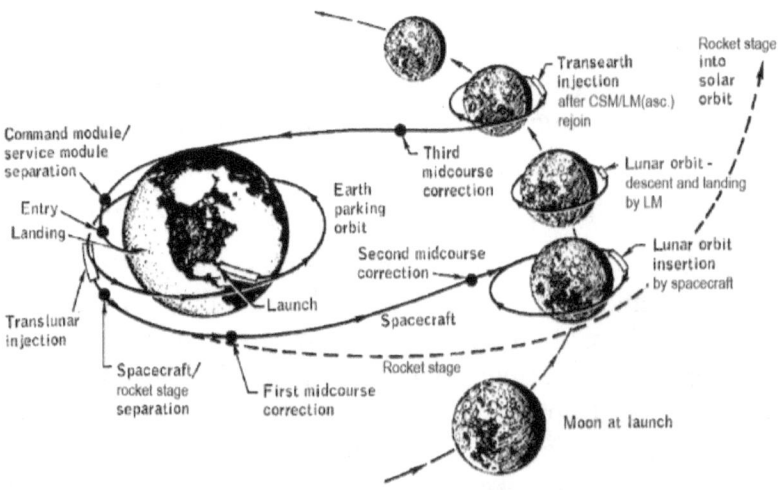

*Figure 1 The Profile for the Apollo Mission Lunar Flight*

Under the technical classification, mission specialists and project leaders discuss the feasibility to apply two types of non-propellant propulsion in current and proposed interplanetary flights. One is passive propulsion and the other is active propulsion. Passive propulsion uses external particles and energy, such as plasma and photons from light source and electromagnet in space, to propel the interstellar vehicle. The solar sail in Figure 2 is an example of passive propulsion. Active propulsion brings propulsive resources carried from the space carrier, such as chemical propellant, ionized gas, viscous fluid liquid hydrant and electricity, to generate inertial forces to push the vehicle in interplanetary exploration. The Magnetoplasmadynamic (MPD) Thrusters and Pulse Inductive Thruster (PIT) are the recent developed examples of active propulsion. In 2004, Gilland, Fiehler and Lyons introduce Magnetoplasmadynamic (MPD) Thrusters and Pulsed Inductive Thruster (PIT) at the second International Energy Conversion Engineering Conference. Using the electromagnetic force to accelerate neutrally charged plasma to high velocities is the basic concept for two thruster designs. The authors announced that the specific impulses of these thrusters could reach 1000 to 10, 0000 seconds. Although the Magnetoplasmadynamic (MPD) Thrusters and Pulse Inductive Thruster (PIT) have high specific impulses for space travel, they only provide little propulsion force at the beginning of the space flight and need heat-resistance materials for designing heating components to handle high temperatures during the operation, which is not efficient for manned interplanetary flights.

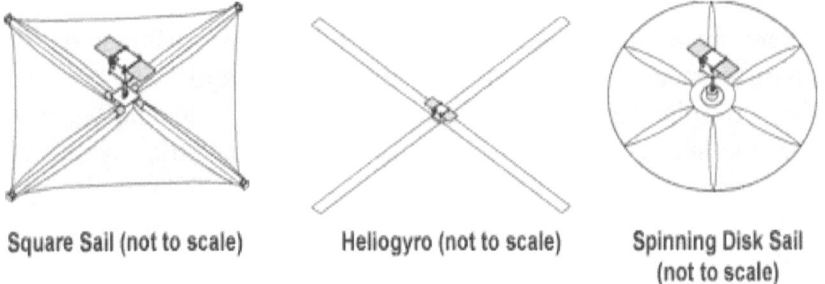

Square Sail (not to scale)    Heliogyro (not to scale)    Spinning Disk Sail
                                                          (not to scale)

*Figure 2 Types of the Solar Sail*

Since the Moon and Mars are two celestial entities for scientific explorations and colonies, engineers and scientists in space agencies must encourage engineers and scientist to develop new propulsive technologies for interstellar vehicles to provide reliable, economical and eco-friendly solutions that would receive needed supports in the public. In order to answer the challenges of space plans, non-propellant propulsion is one of the options to solve unseen barriers in the propulsive system under the space vehicle. In 1998, Mick T. French proposed a new idea of applying non-zero momentum to propel the spacecraft. This is the first conceptual design for non-propellant propulsive disciplines. However, non-propellant propulsion was looked as unrealistic and scientific frictions, not the serious engineering and scientific study and research. Since 2005, academic and industrial scholars have presented several fundamental concepts and theories of non-propellant propulsion in public forums. Fluid viscosity propulsion and impulsive momentum propulsion belong to the classification of active non-propellant propulsion because these two propulsive theories use carbon dioxide from the space vehicle as viscous fluid to generate propulsive forces to the spaceship, and non-zero impulse-momentum theory to move interstellar vehicles under the low gravitational environment in space. These two unique subjects in non-propellant propulsion could solve major challenges, such as the duration of the operation and the reliability in mechanical components in future space transportation.

The purpose for this book introduces new theories and designs based on mathematical formula derived from the fluid viscosity, impulse and momentum in physics. These two design concepts of fluid viscosity propulsion and impulsive momentum propulsion are different from other methods proposed in public discussions. The main difference is that these two designs do not require any propellant during interplanetary flight. These two propulsive technologies will propel the space vehicle to sail in space that has little gravitational effect. The following chapter gives brief descriptions in space environments and orbital transfer methods used for interstellar flights. Chapter 3 presented different types of non-propellant propulsion and two active non-propellant propulsive designs, the fluid viscosity propulsion and the impulsive momentum propulsion published at the AIAA/ASME/SAE/ASEE Joint Propulsion Conferences. Chapter Four evaluates two studies of active non-propellant propulsion for interplanetary transportation and advantages to these two conceptual designs. Finally, the author would discuss some technical suggestions and potential contributions of these two cases of active non-propellant propulsion-fluid viscosity propulsion and impulsive momentum propulsion at the last chapter in this book.

# Chapter 2

# Basics of Space Environment and Orbital Transfers

Space is an unknown frontier for exploratory adventures. Because of unique conditions other than the Earth environment, transferring interstellar vehicles from the Earth orbit to other planets would face high risks during unpredictable situations. This chapter presents brief descriptions of basic space environments and orbital transfer methods for interplanetary transportation. Both two topics would pave the fundamental knowledge for readers to understand fully concepts and technologies behind non-propellant propulsion for scientific explorations.

## 2.1
## Basics of Space Environment

According to the definition, the boundary of space starts from the edge of the Earth atmosphere, which is 62 miles (100 kilometers) above the sea level based on the Kármán line from the Fédération Aéronautique Internationale (FAI). Before the Space Age, scientists assumed space as a vacuum place without any substance. After various successful unmanned scientific missions, scientists discovered vibrant activities and unusual phenomena in space. The following summarizes concise information of basic space environments in temperature, gravitational force and composition.

Earth has thick atmosphere and liquid water to modulate appropriate temperatures through convections and radiations, so that all life forms are able to survive and habitat on the surface. On the contrary, temperatures in space change dramatically in different locations due to the lack of attractive forces to hold gas for heat convections. According to the NASA's fact sheet of space, a bulky white spacesuit would experience 275 degree Fahrenheit differences between one side facing the sun and the other side facing toward the deep space. Moreover, the International Space Station (ISS) installs blankets on important scientific equipment exposed in space to maintain the touch temperature between 120 to -129 degree Celsius. In the deep space far away from the sun and stars, the average temperature is 2.725 degree Kelvin (-120 degree Celsius or -455 degree Fahrenheit) caused all modules to stop moving. However, this extremely low temperature is ideal for the superconductivity that could apply in the energy storage and device to generate electromagnetic force for impulsive momentum propulsion inside the interstellar spacecraft.

Gravitational force is a normal force pointed toward the center of the planet. The gravity is an acceleration to pull all objects to stick firmly on the surface. The higher altitude would have the lower gravity. Equation (1) shows the relationship between the gravity and the altitude:

$$g = g_0 \left( \frac{r_{planet}}{r_{planet}+h} \right)^2 \qquad (1)$$

The standard gravity is a constant that value is at a particular location in the planetary body. For example, the standard gravity on Earth is 9.80665 m/s$^2$ or 32.174 ft/s$^2$ on the geodetic latitude of 45 degrees at the sea level. Each astrological entity has its own gravitational field and force, and always interacts with other gravitational fields from another celestial body. In some particular points at space, the interactions of gravitational fields are balanced. These particular points in space are Lagrangian points, which are ideal to set space stations for the step stone in the planetary exploration. The gravitational force is extremely important to interplanetary transportation because this force will assist the interstellar vehicle to travel other planet through the orbital transfer methods.

Before human beings were able to reach the edge of the atmosphere, scientific fiction imagined space as a giant vacuum without any gas and substance. During investigations of the

International Geographic Year (IGY) in 1957, scientists discovered that space has abundant hydrogen and helium that are intergalactic row materials in the fusion process for the sun and stars, plasma or solar wind ejected from the sun and dark matter through data transmitted from satellites and probes. Because of low attractive forces to hold gaseous atoms, space has rarefiable hydrogen and helium. However, some places near celestial bodies have thick layer of gas that could play the important role in the astrodynamic deceleration for interstellar vehicles. These astonishing discoveries would raise hopes to develop unique propulsion under low gravitational environments to propel interplanetary spaceships by interacting space compositions. For example, ionized particles and plasma ejected from the sun could use as media for electromagnetic propulsion and solar sail, one of passive non-propellant propulsion in the astronautic flight. The concentrated space gas could apply as viscous fluid in fluid viscosity propulsion, one of active non-propellant propulsion for interplanetary transportation. Even dark matter and anti-matter could use in the advanced nuclear propulsive device for fast-speed space vehicles. Composition in space that currently discovered would become advantages of practical applications for future space propulsion in the contribution of planetary explorations.

## 2.2
## Orbital Transfers

The solar system has celestial bodies with various sizes and masses to revolve with designated orbits about the center of the sun. According to the definition, celestial mechanics is the branch of mechanics concerned with the motion of natural or artificial celestial bodies under the influence of gravity. Usually celestial mechanics and orbital mechanics are synonymous and exchangeable. However, academic scholars often use orbital mechanics to describe the trajectories and orbits for human-made vehicles. Although the Newton's three-law of motion is the foundation of orbital mechanics for two bodies interfering in the universe, the Keplerian three-law of planetary motion developed by Johannes Kepler (1571-1630) is always applied in the design of interplanetary transportation during space explorations, which states as follows:

1.  The orbits of the planets are ellipses with the Sun at one focus.

2.  The line joining a planet to the Sun sweeps out equal areas in equal intervals of time.

3.  The square of the period of a planet is proportional to the cube of the major axis of its elliptical orbit.

In standard procedures of interplanetary flights, an interstellar vehicle is on the parking orbit, which is above the average level of the planet's surface for the initial velocity that has the same planetary escape velocity. Then a vehicle transfers from the injection point at the parking orbit to the arrival point at the orbit on the target planet with an additional velocity generating from the vehicle's propulsive system. With useful assistance of the gravitational force from a planet and propulsive engines, a spacecraft could transfer mission crews and cargos from a leaving planet to the explored planet with designated elliptical-shape orbit inside the solar system. In practical applications of interplanetary transportation, the Hohmann transfer ellipse and fast transfer are two methods commonly used in orbital transfers. Although Dr. Buzz Aldrin, the second astronaut to walk on the moon, proposed his orbital transfer method "Aldrin Mars Cycler" to reduce the transfer time and propellant from Earth to Mars at the AIAA Space 2007 Conference, all fields of study in orbital transfers are based on the assistance of gravitational force in each celestial body and propulsion devices. This book gives brief descriptions on both the Hohmann and fast transfer methods without complicated mathematical formula and elaborates theories to summarize the foundation and advantages for space travel.

In 1925, Dr. Walter Hohmann described his innovative orbital transfer technique called "the Hohmann Transfer Method", which paved the way to transport probes and spacecraft from Earth to other planets during the Space Age. The Hohmann Transfer method, which applies a basic transfer trajectory between two circular orbits, is the elliptical shape of the orbit with the perigee at the inner orbit, and the apogee at the outer orbit. During the procedure of a Hohmann

Transfer from an inner circular orbit to an outer circular orbit, this process applies two impulsive velocities generated from propulsive systems. The first impulsive velocity firing a collinear at perigee increases the energy of interstellar vehicle to match the transfer ellipse. Before the velocity reaches the target planet, the second impulsive velocity with a collinear firing at apogee further increases the energy to match the outer orbit. Without the second impulsive velocity, the spacecraft would remain in the transfer ellipse, return to the inner orbit, and continue to orbit the Earth in the elliptical transfer orbit. Figure 3 illustrates the Hohmann Transfer Method to transport an interplanetary carrier from Earth to Mars. Because the Hohmann Transfer has a minimal semi major axis of the elliptical orbit to use minimum energy, which is the significant advantage for planetary explorations, a spacecraft could carry less propellant to reduce the total weight or increase the cargo load during the interplanetary flight. However, this orbital transfer method takes a longer transfer time from one planet to another celestial body. This orbital transfer always applies in unmanned cargo ship and probes because harmful environments exposed in space for a long duration of time do not affect these types of unmanned objects.

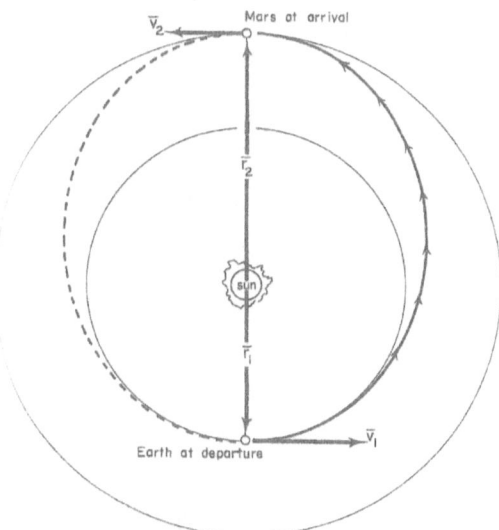

*Figure 3 Hohmann Transfer Orbit to Mars with Three Body System*

On the other hand, the fast transfer method is a different concept compared to the Hohmann Transfer method. A fast transfer could be elliptical in usual cases, parabolic or hyperbolic orbits with at least one noncollinear impulsive velocity. In traditional procedures of a fast transfer method, an interstellar vehicle provides a collinear impulsive velocity at the perigee of an inner orbit. Then a vehicle produces a noncollinear impulsive velocity at the interception of outer orbit. Figure 4 shows the Fast Transfer Method to toss a spacecraft from Earth to Mars. Because a fast transfer has a greater semi major axis of the orbit than the Hohmann Transfer, a space vehicle could spend less time to reach the target of the planet to explore that is beneficial for astronauts to reduce exposing time under the harmful radiation in space. However, this orbital transfer technique needs to use much propellant that dramatically increases the total weight of the interstellar vehicle, or reduce carrying the loads inside the vehicle. Advanced and unique propulsion, such as nuclear propulsion and non-propellant propulsion introduced in the following chapter would be the opinion to generate much needed impulsive velocity on the Earth orbit for the fast transfer method in interplanetary transportation.

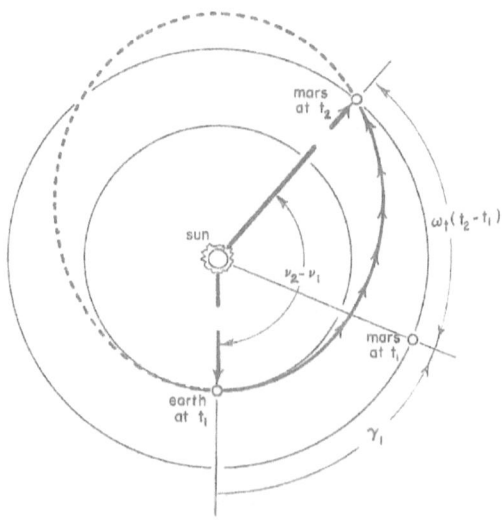

*Figure 4 Fast Transfer Orbit to Mars with Three Body System*

# Chapter 3

## Non-Propellant Propulsive Technologies for Interplanetary Flights

Chemical-type propulsion limits the thrust range to propel giant manned space vehicles from the Earth orbit far beyond the moon. In order to extend scientific explorations to Mars and other planets in the solar system, developing new generation space propulsion becomes the priority for interplanetary transportation. Although scientists and engineers in the space engineering community propose creative ideas in the propulsive field, these concepts are far from the fruition and impractical to integrate into the current design of interstellar vehicles for scientific explorations in other planets. The following sections of this chapter gives brief introductions of different types of non-propellant propulsive theories, and two non-propellant propulsion, viscosity propulsion and impulsive momentum propulsion presented at the AIAA/ASME/SAE/ASEE Joint Propulsion Conferences, for feasible studies in space propulsion under the manned interstellar craft to contribute scientific planetary explorations.

## 3.1
## Types of Non-Propellant Propulsive Concepts

Non-propellant propulsion has different concepts of physics compared to other conventional propulsion subjects. One significant difference is no mass ejection from the interstellar vehicle that the interplanetary vehicle could eliminate the necessary to carry any fuel or propellant. Because non-propellant propulsion uses the space-time medium as the energy resource or working fluid around the vehicle, the specific vacuum impulse (Isp) is unavailable to evaluate the performance. In space, non-propellant propulsion relies on advanced concepts in fluidic space-time, quantum physics, string theory, electromagnetism and gravity to create possible propulsive forces for space transportation. Some employ aspects of cosmological genres, such as dark matter, dark energy, black holes, gravity wave, alternative dimensions and universal expansion are highly speculative, but have strong foundations rooted in current scientific knowledge and experimental observations. These proposals could not only propel a vehicle at very high sub light speeds, which are from seventy percent to ninety percent of light speed, but at the speed of light or beyond! This advantage is necessary for truly reasonable, manned interstellar missions in the solar system or between two galaxy systems!

In categories of unique propulsion for space transportation, various scholars presented three major types of non-propellant propulsion in public forums. The first type is the space-time warp system, which modifies the space-time continuum to mitigate relativistic effects and allow for travel. The Alcubierre Warp drive and traversable Wormhole are two examples for this type of propulsion. The second type is the gravity, inertia and electromagnetic coupling, which mitigates, reduces or artificially creates gravity and inertia propulsive forces through novel electromagnetic interactions, such as the Grand Unified Theory. Heim Theory, Gravito-Electromagnetism (GEM), and Mach's Principle and Mass Fluctuations are excellent examples for this propulsion. The last type is the alternative dimension and hyperspace that enters an alternative space-time where relativistic effects are circumvented and faster-than-light travel is natural and possible. Hyperspace in General Relativity, String Theory based on Alcubierre Drive, and Tri-Space and Fluidic Space-Time are famous examples for this propulsion. The following paragraphs present brief introductions in Heim Theory, Gravito-Electromagnetism (GEM), and Mach's Principle and Mass Fluctuations for different non-propellant propulsive fields other than fluid viscosity propulsion and impulsive momentum propulsion.

Dr. Burkkard Heim developed the Heim Theory between the 1970s and 1980s as an approach to the "Theory of Everything" that attempts to unify gravitation and quantum mechanics with heavy mathematics. He proposed to convert photons into "gravitio-photons" via quantum hyperspace resulting in a measurable force, which could be used for propulsion. His theory inspired Dr. Walter Dröscher and Dr. Jochem Häuser from Germany to devise several ways to extend Heim Theory into advanced propulsive concepts beginning to gain recognition as a viable alternative to the standard model, modern physics and quantum mechanics. However, Dr. Heim only had one peer-reviewed publication in 1977 and other non-reviewed documents. His predictions of new particles and forces described in his theory, which some particles, and forces that observed do not account, do not yet observe or experimentally verify. Because of no scientific verification of new particles and forces, this propulsive theory is still under the category in scientific fictions.

The theory of the Gravito-Electro Magnetism attempts to merge the equations and explanations involving particle physics and quantum physics with heavy mathematics, such as Einstein Field equations, stress-energy tensors, torsion fields, etc., for gravity and inertia with those of electromagnetism to create or manipulate gravity through precise control of electromagnetic forces. Antennae, coils, toroidal inductors, and other hardware could generate an anti-inertia field as inertia dampers to provide vehicle protection from rapid accelerations. This theory has thoroughly examined, but never successfully demonstrated for over ninety years. Mr. J. Brandenburg from ORBITC actively pursued propulsion applications and theory development for space transportation. Many engineering approaches have existed, and some developments even presented on how to augment or attenuate gravity by controlling electromagnetic fields. However, very few concepts have experimentally tested that reveals with null, unfavorable or questionable results. Even this propulsion is achievable, propulsive performance for speeds at or near the light is unknown. This restricts the further development by applying the Gravito-Electro Magnetism for interplanetary flights.

According to Mach's Principle, accelerating object felt inertia due to the radioactive gravitational effects of the distant matter in the universe. Accelerating masses will have different inertia energy than the one at rest. Because mass energy is not Lorentz invariant changing with respect to observers, how does energy change during the process? In addition, when is the resistive inertia forces accounted for? If the scientific equipment could fluctuate to the mass of an object rapidly, the time-averaged pulling reactions on the object due to Mach's Principle may result in a directional force. Using electromagnetic fields, such as alternating the ions in charged capacitors and subjecting them to an oscillating magnetic field, to create rapidly mass fluctuations with the distant matter in the universe will react upon to develop a net force for the space vehicle. Dr. James Woodward from California State University at Fullerton, and Mr. Paul March retired from Lockheed Martin Corporation collaborated with experiments in progress and received favorable results through simple hardware with reasonable power requirements in the laboratory. Through the technical view in the experiments, laboratory Mach-Lorentz Thrusters (MLTs) have produced up to one hundred nano Newton compared to small electric thrusters, but measurements need to be refined. They also predicted that large-scale Mach-Lorentz Thrusters (MLTs) or unidirectional Force Generators (UFGs) might also produce negative mass required for other non-propellant concepts. However, the fundamental theory of Mach's Principle is difficult to understand. In addition, researchers are difficult to measure and quantify mass fluctuation effects due to signal noise contamination and balance effects. In addition, various tests around research and academic institutions do not yet quantify propulsion performances of Mach-Lorentz Thrusters (MLTs). This thruster may only be capable of sub light speeds for interstellar vehicles under current discoveries and developments based on experimental results in the laboratory.

## 3.2
## Fluid Viscosity Propulsion

Biological entities, such as astronauts and other living creatures in an interstellar spacecraft, produce carbon dioxide that is a poisonous gas to harm lives in a closed environment. In order to prevent any accident caused by carbon dioxide, interstellar spacecraft must expel carbon dioxide to space that the average temperature is around 2.725 Kelvin (-279 degree Celsius). According to the theory of viscous flow, carbon dioxide will have higher coefficient of viscosity in the high temperature. This special characteristic gives carbon dioxide to propel giant interstellar vehicles with components by applying the theories of rotational kinematics and impulse-momentum. When a rotational component generates tangent force with a perpendicular distance and the contact area, carbon dioxide will move with a constant velocity. The relationship among the tangent force, constant velocity of carbon dioxide, the contact area, coefficient of viscosity and a perpendicular distance could formulate as:

$$F = \frac{\eta A v}{y} \qquad (2)$$

In order to provide tangent force from a rotational component in this propulsion device, the electric motor generates a constant angular velocity to the transfer belt with a large contacted area to carbon dioxide. The relationships among the viscous force, constant velocity of carbon dioxide, the contact area, coefficient of viscosity, a perpendicular distance, and radius and angular velocity of the rotational component would formulate as:

$$F = \frac{2\eta A r \omega}{y} \qquad (3)$$

The interstellar vehicle will rotate when the rotational component inside the propulsion generates rotational momentum. In order to prevent the rotational momentum on the interstellar spacecraft, another rotational device with opposite rotational direction would install to balance the rotational momentum caused by the first rotational device. Because another rotational device with opposite rotational direction creates the same direction of the linear viscous force from Carbon Dioxide, this propulsion device generates twice propulsion force during the operation in space. To provide the velocity for the interplanetary spacecraft in space explorations, this innovative propulsion applies the principles of impulse and linear momentum.

The tangent force generated from a rotational device will transfer its linear momentum to the spacecraft through carbon dioxide-the viscous fluid in space. This technique provides the initial velocity for the interstellar spacecraft. The relationship among the tangent force, mass and velocity of the interstellar vehicle, time to operate the propulsion device, and components in the rotational device from the rest in orbit shows as:

$$\frac{2\eta A r \omega}{y} t = MV \qquad (4)$$

The mathematical model of the velocity for the interstellar vehicle provided from this fluid viscosity propulsive system show as:

$$V = \frac{2\eta A r \omega}{My} t \qquad (5)$$

This propulsion provides a necessary escape velocity for the interstellar vehicle in a short time span. For example, this propulsion will provide 536,000,000 Newton propulsion force and 1,340,000 m/s to propel the 2,000 kilo-gram spacecraft. The propulsive forces produce through a rotational component of 10,000-meter radius, 1,000 rad/s, 10,000 square meters contact area and 0.01meter of the distance from an immobile surface to the device at the rotational component in the flowing chamber under the condition of 200 degree Celsius carbon dioxide in five seconds from the rest on orbit.

This device uses either the direct current or alternative current that is from the power generator or nuclear power reactor to generate rotational speeds with opposite rotational directions to provide the propulsion force twice to the interstellar spacecraft. Figure 5 shows the design concept and force analysis of the device of the fluid viscosity propulsive system based on equation (3). Figure 5 illustrates two rotational components powered by an electric motor connected with gear components generate opposite rotational directions. When the rotational components rotate with the constant angular velocity with opposite directions, the tangent forces generated from rotational components move treated carbon dioxide exhausted from the interstellar vehicle in the distance between the rotational component and stationary plate. Because this fluid viscosity propulsive device has two exits to produce the propulsive force twice, the final propulsive force will be able to propel the giant interstellar spacecraft without any propellant carried in the interplanetary transportation. This conceptual design provides continuous propulsion force to the interstellar spacecraft through a transfer of momentum method interacted with carbon dioxide from the interstellar spacecraft during interplanetary flight.

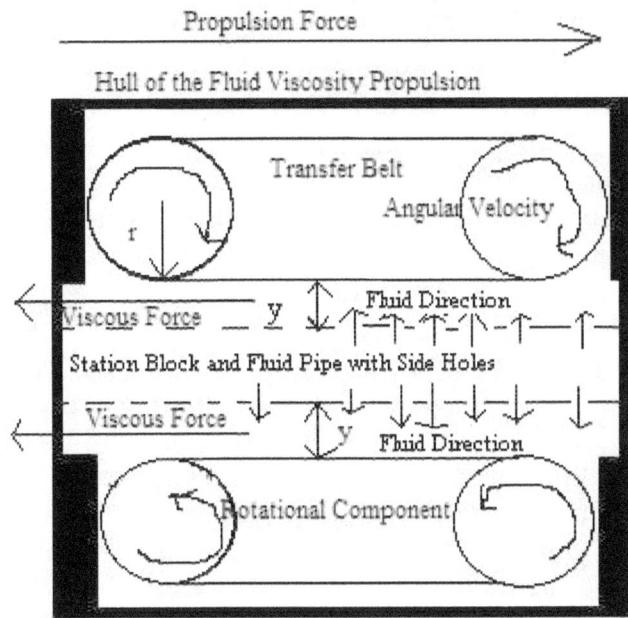

*Figure 5 The Design Concept of the Device of the Fluid Viscosity Propulsion Unit*

# 3.3
# Impulsive Momentum Propulsion

When a charge is in an electric field, that charge experiences an electric force. The magnetic force would generate when met two conditions:

A.    The charge must be moving, for no magnetic force acts on a stationary charge.

B.    The velocity of the moving charge must have a component that is perpendicular to the direction of the magnetic field.

Based on those two conditions, the magnitude of the electromagnetic force is directly proportional to the magnitudes of (1) the charge and (2) the component of the velocity perpendicular to the magnetic field. This is how the electromagnetic force occurred, also known as the Lorenz force. Because of these factors, the relationship between the magnitude of the magnetic field and the velocity of the moving charge formulate as:

$$\vec{F} = q(\vec{v} \times \vec{B}) \text{ or } F = qvB \sin \theta \qquad (6)$$

In equation (6), B is a vector, and its direction can be determined by using a small compass needle to find the magnetic force direction.

From the experiment of the physics of the magnetic field in the electric field, a charge moving through a magnetic field can experience a magnetic force. Since an electric current is a collection of moving charges, a current in the presence of a magnetic field can also experience a magnetic force. In the laboratory experiment, the direction of a current affects the direction of the magnetic force. This phenomenon helps propulsive engineers to apply in the impulsive momentum propulsion design to control the flight directions in interplanetary flight. Support the time required for the charge to travel the length of the wire, equation (6) will be substituted I and $v$ as

$$F = (\frac{q}{t})(vt)B \sin \theta \qquad (7)$$

Because $I = \frac{q}{t}$ and $L_c{=}vt$, equation (7) is formulated as

$$F = IL_cB \sin \theta \qquad (8)$$

Equation (8) shows that the maximum electromagnetic force occurs if the wire is oriented perpendicular to the field ($\theta = 90°$). The new equation from equation (8) will present as

$$F = IL_cB \qquad (9)$$

In the design of the impulsive momentum propulsion system, the solenoid is the best choice to manufacture for this device of impulsive momentum propulsion. The solenoid can gain much electromagnetic force for propulsive forces in the interplanetary flight. A solenoid is a long coil of the wire in the shape of a helix, and the force field provided from the solenoid must move a metal object to transfer momentum. The field inside the solenoid and away from its end is nearly constant in magnitude and directed parallel to the axis. The magnitude of the magnetic field in the interior of long solenoid is

$$B = \mu_0 \frac{n}{L} I \quad (10)$$

Assume the length of the solenoid is

$$L_c = \pi d n \quad (11)$$

When substituting B and $L_c$ into equation (9), the new equation of the electromagnetic force in the solenoid type design will be presented as

$$F' = \frac{2\mu_0 \pi d I^2 n^2}{L} \quad (12)$$

To provide the velocity for the spacecraft in the space exploration, this innovative propulsive technology applies the principles of impulse and linear momentum. The electromagnetic force attracts the moving metal pellets at the end of the platform, where the moving pellets float in the space. The moving iron bar will transfer its momentum to the spacecraft. This technique provides the initial velocity for the spacecraft. The relationships among the electromagnetic force, the moving iron part and the spacecraft from the rest in orbit shows as:

$$Ft = mv = MV \quad (13)$$

The mathematical model of the spacecraft velocity provided from this non-propellant propulsive system shows as

$$V = \frac{2\pi\mu_0 dI^2 n^2 t}{ML} \quad (14)$$

This propulsion provides an initial velocity for the vehicle in a short time span. In order for continued acceleration, the propulsive components inside the platform of a spacecraft would have to rotate with opposite directions, and the magnetic field would have to reverse if changing flight paths. For example, this propulsion will provide 789.57 m/s to propel the 10,000-kilo gram spacecraft with only 10 ampere, one-meter diameter, one-meter length and 100,000 turns of the coil on the solenoid in one second from the rest.

This device uses the direct current to attract the moving metal pellets to provide the initial velocity to the spacecraft. Space pilots must reverse the direction of the direct current for each 180° rotation of the platform in order to continue its acceleration. The direct current is from the power generator that created by a nuclear power source in the spacecraft. Figure 6 illustrates the conceptual design of impulsive momentum propulsion for interplanetary flight. Two solenoids located at the end of the platform. Both two tubes on both sides have receiving devices with sensors for sending signals to two motors to rotate 180° on both tubes. After the solenoid energizes the moving component with the Direct Current, two metal pellets move to the end of tubes directly. After receiving devices catch both two pellets, the sensors send signals to the on-board computer. The on-board computer will terminate the current on solenoids, and activate the motors to reverse both two tubes for 180° rotations. After both two tubes are in place, the impulsive momentum propulsive system repeats the same procedure to contact solenoids to energize the moving metal pellets to attract to the end of the tubes again. This conceptual design provides continuous propulsion to the spacecraft through a transfer of momentum method during interplanetary flights.

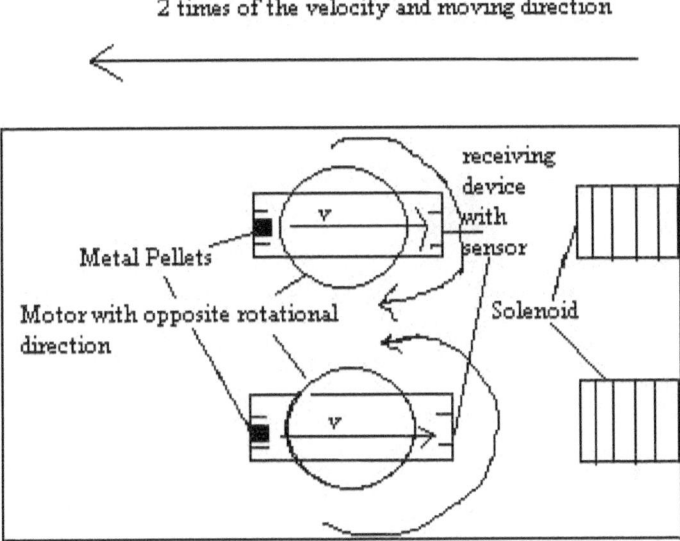

*Figure 6 The Design Concept of the Device of the Impulsive Momentum Propulsion Unit*

# Chapter 4

## Evaluations of Non-Propellant Propulsion for Interplanetary Transportation

Non-propellant propulsion is a unique concept in space propulsion. Three significant advantages are (1) simple to build and easy to maintain, (2) high performance in space propulsion compared to other conventional space propulsive technologies, and (3) the fast orbital transfer method produces enough escape velocities for interplanetary flights. Here are the detailed reasons that choosing non-propellant propulsion is a candidate for interplanetary transportation:

1.  The non-propellant propulsive system has the simple structure based on the discussion of the conceptual design in the previous section. This significant advantage helps construction companies reduce the budget to build and maintain the propulsive system integrated in the interstellar vehicle. In addition, these innovations do not carry any propellant mass. Mission crews will have a safe interplanetary flight because of no dangerous chemical propellant carried inside the spacecraft.

2.  The non-propellant propulsive system performs a wide range of thrust ranges but no specific vacuum impulse compared to conventional and advanced propulsive designs from major aeronautical companies for space flights. The ideal space propulsive system should have a wide range of thrust and highest specific vacuum impulse (Isp) for interplanetary flights. The highest specific vacuum impulse can reach 6,000 seconds from

electrostatic ion propulsion. However, the thrust from this system is too low to propel small-unmanned space probes for planetary missions. Both fluid viscosity propulsion and impulsive momentum propulsion under the category of non-propellant propulsion have no specific vacuum impulses. On the contrary, both fluid viscosity propulsion and impulsive momentum propulsion provide various thrust range, because fluid viscosity propulsion changes the range of the angular velocity inside the device to produce different thrusts, and electromagnetic attractive forces generated from solenoids at the end of the impulsive momentum propulsive device to produce a wide range of propelling forces. Table 1 shows the comparison of performances and operating characteristics for space propulsion system from conventional technologies and this conceptual propulsion.

| Table 1 Comparison of performances and operating characteristics | | | |
|---|---|---|---|
| **Type** | **Energy** | **Vacuum Impulse Specific (Isp)** | **Thrust Range (N)** |
| Solid motor | Chemical | 280-300 | $50 - 5 \times 10^6$ |
| Cold gas | High pressure | 50-75 | 0.05-200 |
| Electro thermal resist jet | Chemical | 150-700 | 0.005-0.5 |
| Arc jet | Resistive heating h~0.9 | 450-1500 | 0.05-5 |
| Electrostatics ion | Electro arc heating h~0.3 | 2000-6000 | $5 \times 10^{-6} - 0.5$ |
| Electromagnet ic MPD | Magnetic | 2000 | 25-200 |
| Pulsed plasma | Magnetic | 1500 | $5 \times 10^{-6} - 0.005$ |
| Pulsed inductive | Magnetic | 2500 | 2-200 |

| Table 1 Comparison of performances and operating characteristics | | | |
|---|---|---|---|
| **Type** | **Energy** | **Vacuum Impulse Specific (Isp)** | **Thrust Range (N)** |
| Fluid viscosity propulsion | Viscous fluid | Unavailable | 0 and above |
| Impulsive momentum propulsion | Magnetic | Unavailable | 0 and above |
| Liquid | Chemical | 300-450 | $5 - 5 \times 10^6$ |

3.  Non-propellant propulsion provides enough escape velocities for the fast orbital transfer to reduce the flight time. The fast orbital transfer method can reduce the traveling time in interplanetary flights. However, the barrier of using this method is that the spacecraft must use much propellant to produce the enough initial velocity to sail other planets. It is not benefit of using the fast orbital transfer method with conventional propulsion systems. The non-propellant propulsive system produces enough thrust and continuous velocity to propel the giant spacecraft, and space pilots can accelerate enough escape velocities in a short time. With this unique propulsion, mission crews can reach to the moon within 24 hours, and within three days to Mars. Table 2 shows the flight time of using fluid viscosity propulsive system to propel the 2,000-kilogram interstellar vehicle in five seconds at the altitude of 500 kilometer on Earth and target planet orbits. The fluid viscosity propulsive system uses 10,000-meter radius, 1,000 rad/s, 10,000 square meters contact area and 0.01meter of the distance from an immobile surface to the device in the rotational component and the $2.68 \times 10^{-5}$ Pa·s coefficient of viscosity for carbon dioxide at the 200 degree Celsius. Table 3 shows the flight time of using impulsive momentum propulsive system to propel the 2,000-kilogram interstellar vehicle in one second at the

altitude of 500 kilometer on Earth and target planet orbits. This propulsive device uses two-meter diameter, one-meter long, 1,000,000 turns of the coil and 20 amps on the system.

| Table 2 Flight time with fluid viscosity propulsion ||
| :---: | :---: |
| **Planet** | **Traveling Time** |
| Moon | 0.0791 hour |
| Mars | 1.4538 days |
| Jupiter | 6.4535 days |
| Saturn | 11.9906 days |
| Uranus | 24.2110 days |
| Neptune | 37.9886 days |
| Pluto | 49.8382 days |

| Table 3 Flight time with impulsive momentum propulsion ||
| :---: | :---: |
| **Planet** | **Traveling Time** |
| Moon | 0.0336 hour |
| Mars | 14.989752 hours |
| Jupiter | 2.77286 days |
| Saturn | 5.15197 days |
| Uranus | 10.4026 days |
| Neptune | 16.3223 days |
| Pluto | 21.4135 days |

# Chapter 5

## Conclusion

The non-propellant propulsive system is a unique discipline for future interplanetary transportation. The previous chapter discussed the outcomes in the developments of non-propellant propulsion-fluid viscosity propulsion and impulsive momentum propulsion. Here are some suggestions of non-propellant propulsion in future space missions.

First, non-propellant propulsion provides no specific impulse in space. Because both two non-propellant propulsive systems generate huge forces and accelerations immediately, these propulsive designs are the best candidates for unmanned space probes and robotic space missions in the near-term space mission. They save time and budget for planetary explorations in the solar system, and fulfill new design requirements in space transportation, such as the safety and lighted but powerful propulsive engines on manned exploration vehicles for long-range space explorations. These propulsive concepts will thus be of value in developing scientific exploration goals for manned exploration vehicles, if on-board computers can modify the suitable rotational velocity in rotational components at the fluid viscosity propulsive equipment, and electromagnetic forces to attract metal pellets at the impulsive momentum propulsive system to produce appropriate thrusts for astronauts to handle accelerations in interplanetary travels.

Next, fluid viscosity propulsive technology has a minor barrier because of the resource of carbon dioxide. The main resource of carbon dioxide comes from the waste of the interstellar vehicle. Before sending carbon dioxide to the propulsive device for propelling the spacecraft, service module in the interstellar vehicle must treat carbon dioxide to reach the optimal coefficient of viscous fluid. However, the interstellar spacecraft could not produce enough carbon dioxide for a long distance interplanetary flight. In order to expand the resource of carbon dioxide to propel the interstellar space vehicle, the unique equipment that collects carbon dioxide from ion particles in space could convert ion gas into viscous fluid to produce much needed carbon dioxide for interplanetary flights. In addition, the computer-controlled system could integrate with the fluid viscosity propulsive system to stabilize the highest quality of carbon dioxide to propel the interplanetary vehicle.

Finally, space pilots can control flight directions and speeds easily during interplanetary flight on both fluid viscosity propulsion and impulsive momentum propulsion. Traditionally, space pilots needed to control burning rates of rockets to keep correct flight paths and directions in conventional interstellar crafts. Both fluid viscosity propulsion and impulsive momentum propulsion under this unique propulsion help space pilots operate the spacecraft without controlling any complicated equipment during interplanetary flights. Space pilots only change rotational velocities to reduce or increase speeds, and modify locations of propulsive devices to change flight paths in fluid viscosity propulsion, and the current in the electromagnetic solenoids for attractive forces at impulsive momentum propulsion. Both two conceptual propulsive technologies are more convenient to use for space missions compared to current and under developing propulsive technologies.

# APPENDIX A

## Nomenclature

| | |
|---|---|
| $A$ | contact area with viscous fluid |
| $y$ | distance from an immobile surface in the device |
| $\omega$ | angular velocity of the rotational component in the device |
| $r$ | radius of the rotational component in the device |
| t | time |
| $\eta$ | coefficient of viscosity |
| $V$ | velocity of the interstellar vehicle |
| $F$ | tangent force |
| $M$ | mass of the interstellar vehicle |
| $B$ | magnetic field |
| $d$ | diameter of solenoid |
| $F'$ | electromagnetic force |
| $I$ | current |
| $L$ | length of the solenoid |
| $Lc$ | length of the coil on the solenoid |
| $n$ | turns of coil |
| q | total charge |
| $v$ | velocity of the iron-moving component inside the device |
| $\theta$ | degree |
| $\mu_0$ | permeability of free space ($4\pi\times10^{-7}$ T*m/A) |
| $h$ | altitude from the sea level of the average surface on the planet |
| $g$ | gravity |
| $g_0$ | standard gravity |
| $r_{planet}$ | mean radius of the planet |

# APPENDIX B

## Planetary Physical Data

| Name | Radius ($\times 10^6$ m) | $\mu$ (m³/s²) | Relative Mass | $V_{esc}$ (m/s) |
|---|---|---|---|---|
| Sun | 696.0 | $1.327 \times 10^{20}$ | 332,900 | 617,500 |
| Moon | 1.738 | $4.903 \times 10^{12}$ | 0.0123 | 2,375 |
| Mercury | 2.439 | $2.168 \times 10^{13}$ | 0.0544 | 4,217 |
| Venus | 6.052 | $3.248 \times 10^{14}$ | 0.8149 | 10,360 |
| Earth | 6.378 | $3.986 \times 10^{14}$ | 1.000 | 11,180 |
| Mars | 3.393 | $4.297 \times 10^{13}$ | 0.1078 | 5,032 |
| Jupiter | 71.40 | $1.267 \times 10^{17}$ | 317.8 | 59,568 |
| Saturn | 60.00 | $3.791 \times 10^{16}$ | 95.11 | 35,550 |
| Uranus | 25.40 | $5.788 \times 10^{15}$ | 14.52 | 21,350 |
| Neptune | 24.30 | $6.832 \times 10^{15}$ | 17.14 | 23,710 |
| Pluto | 1.500 | $9.965 \times 10^{11}$ | 0.0025 (?) | 1,153 |

PS: (1) The mass of the Sun is $1.989 \times 10^{30}$ Kilogram.
    (2) The mass of the Earth is $5.975 \times 10^{24}$ Kilogram.

# APPENDIX C

## Planetary Orbital Data

| Name | Semi major Axis (A.U.) | Orbital Velocity (m/s) | Period (days) | Eccentricity ($\varepsilon$) | Inclination (degree) |
|---|---|---|---|---|---|
| Mercury | 0.3871 | 47,870 | 87.97 | 0.2056 | 7.004 |
| Venus | 0.7233 | 35,040 | 224.7 | 0.0068 | 3.394 |
| Earth | 1.000 | 29,790 | 365.2 | 0.0167 | 0.000 |
| Mars | 1.524 | 24,140 | 687.0 | 0.0934 | 1.850 |
| Jupiter | 5.203 | 13,060 | 4,332 | 0.0482 | 1.306 |
| Saturn | 9.539 | 9,650 | 10,760 | 0.0539 | 2.489 |
| Uranus | 19.18 | 6,800 | 30,690 | 0.0471 | 0.773 |
| Neptune | 30.07 | 5,490 | 60,190 | 0.0050 | 1.773 |
| Pluto | 39.44 | 4,740 | 90,460 | 0.2583 | 17.14 |

PS: One Semi major Axis (A.U.) is $1.496 \times 10^{11}$ meters.

# APPENDIX D

## References

NASA Breakthrough Propulsion Physics (BPP) Project," URL: http://www.grc.nasa.gov/WWW/bpp/

Lyons, V., Gilland, J., and Fiehler, D., "Electric Propulsion Concepts Enabled by High Power Systems for Space Exploration," 2nd International Energy Conversion Engineering Conference, Providence, Rhode Island, 2004.

Sun, Y., "Design Concept of Electromagnetic Propulsion for Interplanetary Flight," 41st AIAA/ASME/SAE/ASEE Joint Propulsion Conference, Tucson, Arizona, 2005.

Meholic, G., "Advanced Space Propulsion Concepts for Interstellar Travel," AIAA Los Angles Dinner Meeting, Los Angeles, California, 2008.

Hale, Francis J., Introduction to Space Flight, Prentice-Hall, Inc., 1994.

# Index

www.ingramcontent.com/pod-product-compliance
Lightning Source LLC
Chambersburg PA
CBHW021934170526
45157CB00005B/2314